To Luciano De Crescenzo,
writer and engineer,
for fortifying in me
the extraordinary alchemy
of literature and engineering,
of mythology and science.

Dionigi Cristian Lentini

PraiseENG

A praise of the Engineer

Original Version (Italian): PraisING – Elogio dell'Ingegnere (2018)

Traduction by: Giovanni Frosio

Editor: Tektime

Scientists dream about doing great things. Engineers do them.

(James Michener)

It is told that Pallas Athena, the goddess of wisdom and

beloved daughter of Zeus,

was gifted by her father the clay of Apollon and Dionysus,

the same ones with which men were created,

and on a summer day, in her generous inspiration,

she kneaded out of them a fistful of science and one of technique:

thus was born *ingenium*, the Latin word from which derive "ingenuity",

"engineer", "engine", "ingénieur", "ingegno", "ingegnere ", "ingegneria", etc.

Since then man has wondered about being, becoming,

the immanent and the transcendent,

he has looked into the tenth of millionth of billionth of centimeter of the proton's structure

as well as into the thousands of trillions of kilometers

of the most remote galaxies in the universe;

he has computed the infinitesimal fraction of a second

as well as the lifespan of stars;

he has travelled in many space and time contexts,

leaving behind him a trail of destruction, wars and death,

but also of achievements, hopes and life;

he has crossed the centuries as the undisputed protagonist of History

overbearingly showing his greatest strength in his thoughts,

his most lethal weapon, the feature which sets him apart from any other living creature;

but it's in the subsequent step, of translating thought in innovations

by means of a project,

that he made possible the progress of his kind,

making engineering the natural field of convergence

of science, mathematics and technology.

Engineering is not a mere theoretical speculation

but it's a primary conquest of mankind that, thanks to it,

starting from an analysis or a requirement, a need,

in the realization of physical, social, political and economic constraints,

has been able to provide methodological and design solutions

to supply, control, develop and manage well a material or immaterial commodity,

whether it is a product, a system, a service, a process, an organization.

Engineering is not only a history of human conquers and innovations,

of proactive, revolutionary and futuristic ingenious ideas,

but it's what has allowed man to rule changes,

without passively undergo them, to adapt to them and thus to survive.

Without it, mankind would be an already extinguished species.

Engineering is not only an extraordinary capacity of interpreting human needs

and applying scientific findings in everyday life

but, in a continuous one-to-one interface,

it has affected and change the way, style and philosophy of life,

the effectiveness and efficiency of scientific research,

feeding with its propulsive force

the uninterrupted path of mankind towards progress.

And the Promethean mediator between knowledge and need, between science and society,

is the engineer.

From the agri-food to the biomedical sector,

from civil to management,

from forensic to aerospace,

from mechanical to electronic,

from informatic to nuclear,

from transports to industrial,

from naval to electric, ...

in any sector of application and human organization

engineers have allowed our kind

to cross the centuries of its own existence as a protagonist,

from prehistory to the extraordinary and fascinating era we live in.

From the Mesopotamian wheel to the airplane,

from the pyramids of Ancient Egypt to Burj Khalifa,

from Stonehenge megaliths to atomic clocks,

from the lighthouse of Alexandria to the radar,

from the abacus to the microchip,

from the lightbulb to the transistor,

from the Alcántara to the Golden Gate bridge,

from the Great Wall to the space shield,

from the hydraulic screw to the Panama Canal,

from the armillary sphere to the orbiting spatial station,

from the consular roads to the trans-Siberian,

from the stone arc to the reinforced concrete,

from Leonardo's codes to the cryptographic ones,

from the electric battery to the particle accelerator,

from the Amerigo Vespucci to Internet,

from the basilica of Santa Sofia to the Cathedral of Florence,

from the parachute to Sputnik 1,

from the telescope to the space probes,

from the Millau viaduct to the Palm Islands,

from the Pascaline to the personal computer,

from the pacemaker to the CAT,

from the six-shot pistol to the atomic bomb,

from the telephone to the world wide web,

from the radio to the television,

from the piano to the digital audio,

from the movable type to the 3D printing,

from the cross-bow to the laser,

from plastic to optical fibers,

from the pressure pot to the fridge,

from the temples of Agrigento to the social networks,

from the Nautilus to the Channel,

from Enigma to the RSA algorithm,

from the Panzer to the robots,

from the aqueduct of Segovia to the bridge of Brooklyn,

from the Suez Canal to the Hoover dam,

from the cloning to the coding of the genome,

etc. etc.,

an endless roundup of extraordinary wonders and phenomenal innovations

that the media often abstractly label as "engineering miracles"

but in fact are concretely given birth by the ingenuity of men and women,

single or associated, whose names are eternally bound to burn

to enlighten Mankind:

Lu Ban, Archimedes of Syracuse, Hero of Alexandria,

Filippo Brunelleschi, Leonardo da Vinci, James Watt,

Carlo Vanvitelli, Agustín de Betancourt, George Stephenson,

Nicolas Léonard Sadi Carnot, Isambard Brunel, Mr and Mrs Roebling,

Henry Bessemer, Eugenio Barsanti and Felice Matteucci,

Nikolaus August Otto, Gustave Eiffel, Emil Winkler,

Galileo Ferraris, Thomas Edison, Alexander Bell,

Hertha Ayrton, Nikola Tesla, Rudolf Diesel,

Henry Ford, the Wright brothers, Mr and Mrs Gilbreth,

Nath Khosla, Adriano Olivetti, John von Neumann,

Shigeo Shingo, Fritz Leonhardt, Konrad Zuse,

Hans von Ohain, Wernher von Braun, Hedy Lamarr,

Charles Bachman, Fazlur Khan, Neil Armstrong,

Michael Bloomberg, Jeff Bezos, Larry Page,

...

... it's impossible to mention them all,

in every century or era of history,

in every nation or place on Earth,

their name,

honor and burden at the same time,

is ENGINEER.

According to Le Corbusier

the engineer is he who "inspired by the economic law and guided by computation,

mette in comunicazione armonica l'uomo con le leggi dell'universo";

according to Herbert Hoover it's him to whom

"falls the job of clothing the bare bones of science with life, comfort and hope";

for his detractors, instead,

he's a scruffy nerd, learned, intelligent but careless of his physical look,

an anti-esthete, nagging, pedantic, touchy, at times exalted and superb,

constantly projected into the future, unable to live the present

because of his limited interpersonal skills;

for others yet

he's a crazy and hungry technologist

with a compelling need to put his knowledge and studies to good use,

a futurist and enthusiastic optimist who can only see challenges and opportunities

where others only see problems,

or more simply ...

... an eternal student who's woken up as a man

with an irrepressible need to believe

that something mad, untried and extraordinary

is really possible.

PraiseENG

A praise of the Engineer

The author

Dionigi Cristian Lentini, engineer, professor, science communicator and writer, is born in Mottola in 1980. In the small Apulian town, also known as "the spy of the Ionian Sea", he attends the scientific high school "A. Einstein", where, besides training in the logical - scientific - technological field, he cultivates the humanistic passion for Literature and in particular History and Philosophy. In this period he also begins writing poems and aphorisms. In 1998 he ranks second at the National Prize *Ori di Taranto* - 6th Edition and in the same year he's a finalist at the European Competition *"Who has the right to human rights?"*.

In 2005 he graduates in **Informatic Engineering** at the University of Calabria (*Università degli Studi della Calabria*) with an experimental thesis on "wireless networks security" and before finishing his master's degree course he's already collaborating with the Department of Electronics, Informatic and System Science (D.E.I.S.) of UNICAL and with the CERN of Geneve in the scientific research field; in this context in 2006 he releases the paper *"Static and Dynamic 4-Way Handshake Solutions to Avoid Denial of Service Attack in Wi-Fi Protected Access and IEEE 802.11i"*, reviewed on the EURASIP Journal on Wireless Communications and Networking (JWCN 2006). Still in 2006 he gets the qualification to practice the **engineering profession** and he's listed in the Main Register of the Order of Engineers of the Province of Taranto in the three sectors of

Civil-Environmental, Industrial and Informatic. During the years he'll have an active role in different Commissions of his professional Order and in various stages of the Order's activity.

Fond of mythology and history (in particular Ancient-Medieval and Risorgimento History) from 1994 to 2009 he writes - Boopen Editore - 2005 (his *"Dantis iter inter deum"* among the Greats of History), some accurate cartographic collections (*"Political cartography of Italy"*, *"The Great Empires"*) and then a series of essay compendia and chronologies (*"Chronological summary of historical events concerning the period between the birth of Mesopotamian civilization and the fall of the Western Roman Empire", "Chronological summary of historical events concerning the period between the Barbarian invasions and the discovery of the New World", "Chronological summary of historical events concerning the period between the discovery of the New World e and the fall of Napoleon", "Chronological summary of historical events concerning the period between the Congress of Vienna and the age of Imperialism", "Chronological summary of historical events concerning the period between World War I and 2000", "History of Italy (800 B.C.-1815 A.D.): from Greek colonization to the Congress of Vienna", " History of Italy (1815 A.D.-1861 A.D.): from the age of Restoration to the Unity", " History of Italy (1861 A.D..-1994 A.D.): from the Unity to the Second Republic", "Res Populi Romani", "Eternal Hellas", "Egypt", "The irresistible charm of Ancient East", "Alphabeti", "Calendars", "State Leaders", "Pontifex", "Wars of all times", "Mos Maiorum", History of Sciences", "History of Mathematics", "History of Medicine", "History of Nuclear Physics", "Universal Mythology", "Flags of the world ", "Chronology of art history* ") later collected in *"Annales Rerum Orbis"*.

From 2005 to 2006 he works in Rome as Analyst Consultant at Capgemini Italia S.p.A.- **Capgemini** Group, a multinational working in the field of IT consulting, outsourcing and services; in this short but intense experience he has the chance to participate in various projects in the hi-tech, defense and aerospace fields, in collaboration with Galileo Avionica, Alenia Aeronautica, BAI, etc., and to practice in the field of web-applications, designing and developing the first home-banking platforms for various financial stakeholders (BPU, etc.).

In October 2006 he joins the **Finmeccanica** Group (Leonardo S.p.A., formerly Finmeccanica S.p.A., formerly Alenia Aermacchi S.p.A., formerly Alenia Aeronautica S.p.A, formerly Alenia Composite S.p.A.) at first as Technologic Innovation, Research and Development Engineer (R&D Engineer) and then as ICT Engineer for the international aeronautical program **B-787 (Boeing Dreamliner Program)** at the production plant of Grottaglie (TA).

In 2010 he gets a **Master in General Management** *"BEST – Business Education Strategic Ten"* - ASFOR certified (Corporate Specialized Master) and he writes articles for various magazines in the IT sector such as "SecurInfo" and "Hacker Journal", the firs Italian hacking magazine.

In 2011 he's **expert commissioner** (IT sector) for the Local Action Group *"Luoghi del Mito"* within a rural development plan of the Apulia Region, **external expert** on new technologies within a EU funded national program for various schools and **expert teacher** for various professional formation courses organized by the Orders of Engineers of the Province of Taranto. Since the same year he's technical consultant and expert at the Tribunal for civil and criminal cases, with a specialization in IT Security and Digital Forensics.

Remaining in the IT field, in 2009 he focuses mainly on **web-mastering** (more than 40 websites and web portals designed), **web and mobile applications** (Android), all the while being passionate about cryptography and security protocols.
In 2016 he releases *"Wired and wireless technologies: security protocols of communication networks"* - Lulu Press, *"Cryptography at*

the base of modern technologies" - Lulu Press and "*Hacking Technology - Technologies and Hackers*" - Lulu Press.

With the same editor in 2016 he publishes the poems of his high school and university years in the e-book "*Come una farfalla*", issued also in the English translation "*Like a butterfly*". In 2018 it's the turn of "*9 billion oscillations of cesium*", the second collection of his early poems.

Since 2017 he's **full professor of Technology** at the Ministry of Instruction, University and Research, and **President of the ICT Commission of the Order of Engineers of the Province of Taranto**.

Always attracted by all games with a strong component of reflection and strategy, Cristian sees life as a tough chess game with fate: while the hands of the clock mark the passing of time, it's absolutely forbidden to surrender even one moment, forbidden to give up, forbidden to delude oneself... engage and go on until the checkmate!

His favorite aphorism? "*The world belongs to the enthusiast who keeps cool*" (McFee).

Bibliography

- *Wikipedia, the free encyclopedia. Keyword: Ingegneria (https://it.wikipedia.org/wiki/Ingegneria).*

- *Herbert Hoover, L'elogio dell'Ingegnere - P. 209-210 ; 27 cm. - La Zagaglia : rassegna di scienze, lettere ed arti, A. VIII, n. 30 (giugno 1966)*

- *Aphorisms (free searches on Google.it)*

Being an engineer is actually a disease.
A woman, married to an engineer, could be asked:
"Lady, how's your husband? Is he still engineer?".
Then she could answer: "No, he's slightly better now".

(Luciano De Crescenzo)

www.ingramcontent.com/pod-product-compliance
Ingram Content Group UK Ltd.
Pitfield, Milton Keynes, MK11 3LW, UK
UKHW021926190726
13853UKWH00002B/868

9 788835 417484